U0313062

设计有约 3

inSIDE deSign

香港方黄建筑师事务所商业空间设计系列专集 方峻 著

華中科技大學出版社
http://www.hustp.com
中国 · 武汉

in Hong Kong, China. studied the bachelor/postgraduate/doctor programs and degrees on philosophy, construction engineering and design management in Inter American University, Politecnico di Milano, Huaqiao University, and Hong Kong Polytechnic University. He is a Professional Associations of Hong Kong Interior Design Association (PM00408), Member of International Federation of Interior Architects/ Interior Designers (0281), Institute of Interior Design of Architectural Society of China (8030), Lighting designer of register of China.

The works not only was awarded the golden award at Indoor Design Contest for the 1st International Building Landscape, also was awarded by awards and special honors at indoor design contests home and abroad.

Be more have been inserted by American *Indoor Design* and various well-known magazines.

Here are published and highly acclaimed industry of personal album design have been *Inspiration Design*, *Inside Design I*, *Inside Design II*...*Inside Design V*, *Mangement System and Application for Installation Art Project* etc.

方峻（TFong）

建筑空间与多元跨界的中国香港设计师。

先后在美国美联大学、意大利米兰理工学院、香港理工大学、国立华侨大学

接受哲学、建筑设计、设计管理的学士 / 硕士 / 博士等教育，

同时也是香港室内设计协会专业会员、国际室内建筑师设计联盟会员、中国建筑学会室内设计分会会员、中国注册照明设计师。

其作品不但荣获首届国际建筑景观室内设计大奖赛金奖，

还获得过多项国内外室内设计的奖项与荣誉；更被美国《室内设计》和

各类知名专业杂志数次刊载。相继出版且备受业界好评的

个人设计专辑分别有《“悟”设计》《设计有约 1》《设计有约 2》……《设计有约 5》

《装置艺术项目管理体系与应用》。

Contents
目 录

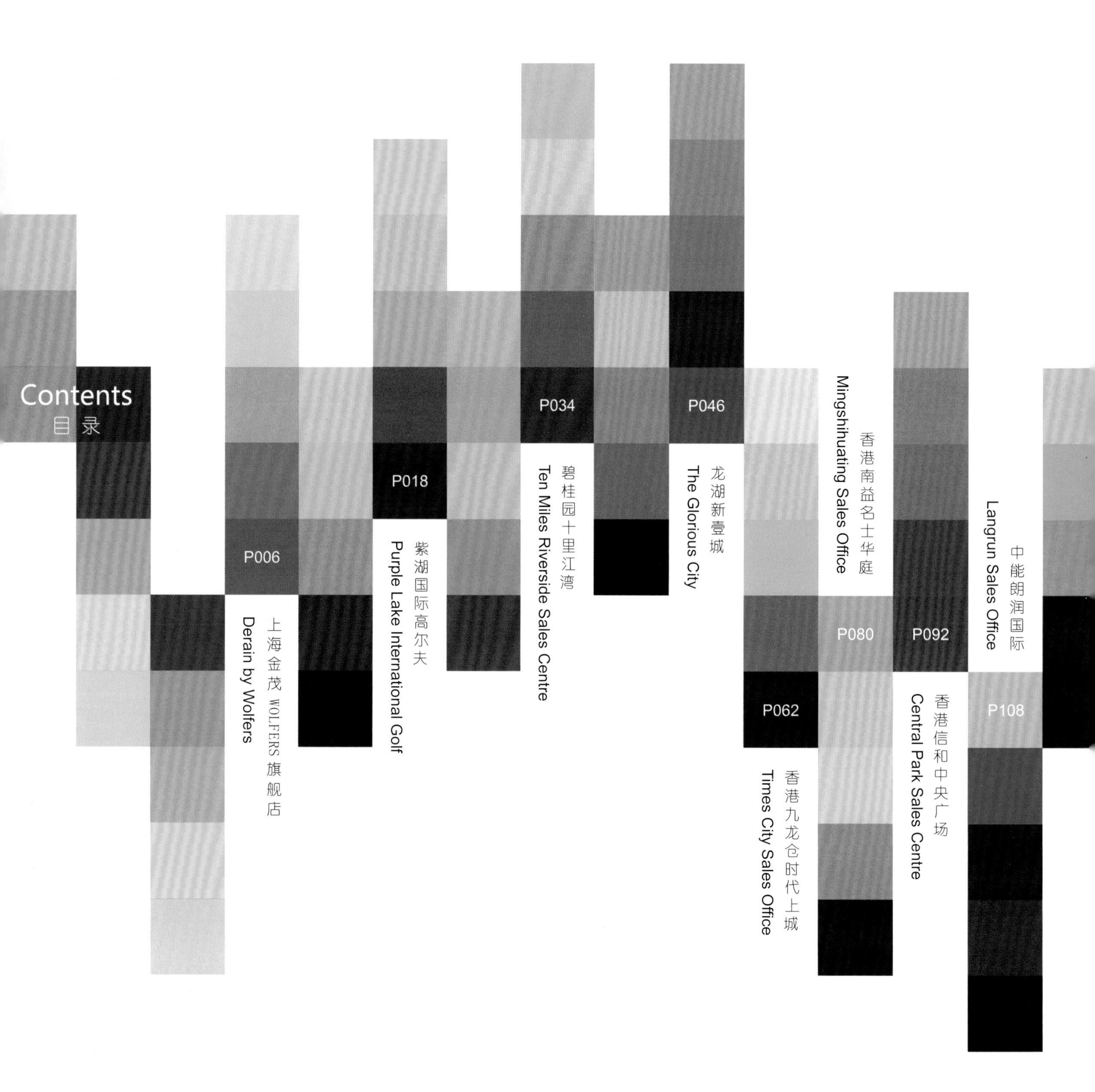

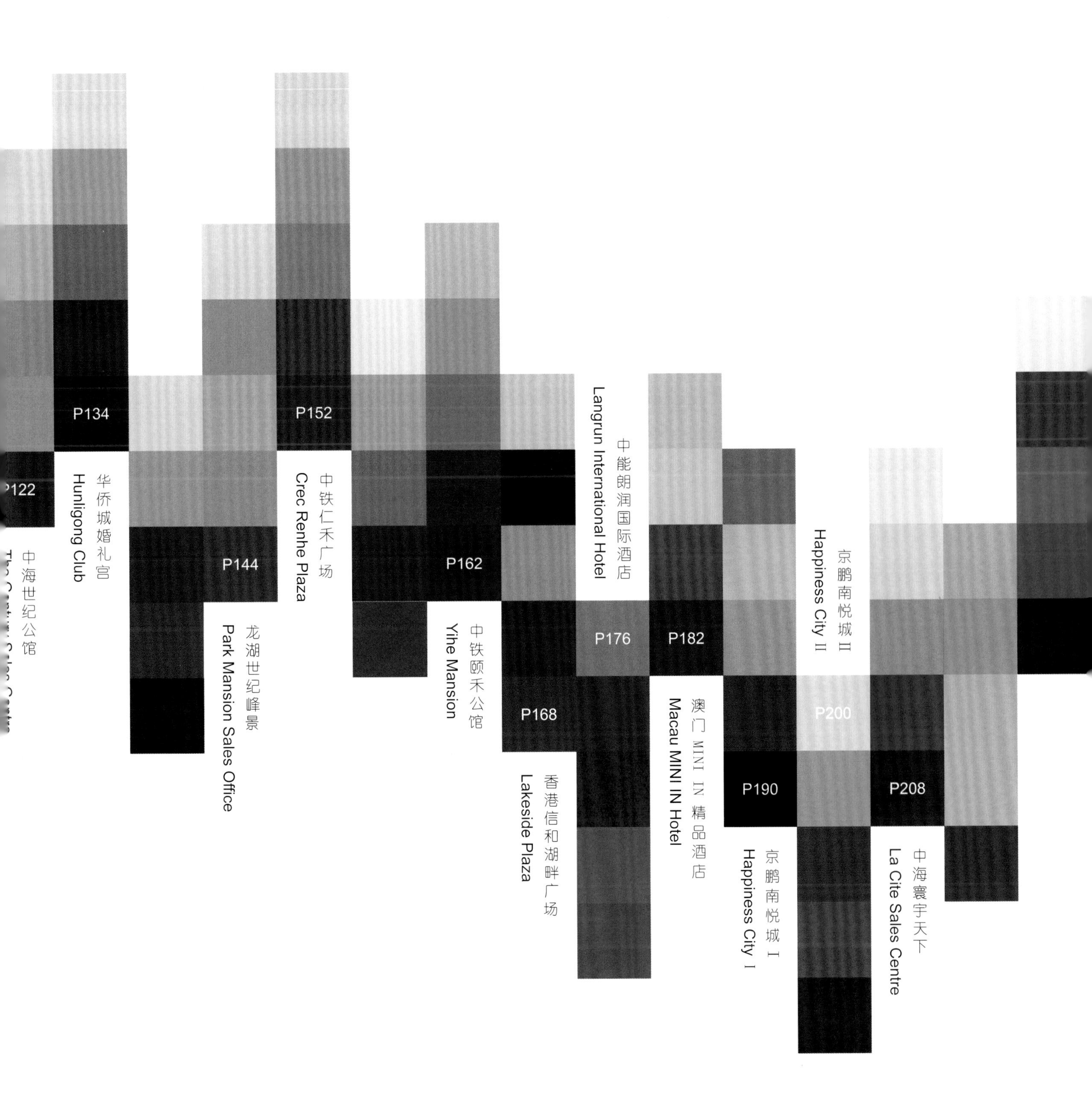

122
中海世纪公馆
P134
华侨城婚礼宫
Hunligong Club
P144
龙湖世纪峰景
Park Mansion Sales Office
P152
中铁仁禾广场
Crec Renhe Plaza
P162
中铁颐禾公馆
Yihe Mansion
P168
香港信和湖畔广场
Lakeside Plaza
P176
中能朗润国际酒店
Langrun International Hotel
P182
澳门 MINI IN 精品酒店
Macau MINI IN Hotel
P190
京鹏南悦城 I
Happiness City I
P200
京鹏南悦城 II
Happiness City II
P208
中海寰宇天下
La Cite Sales Centre

Derain by Wolfers

上海金茂 WOLFERS 旗舰店

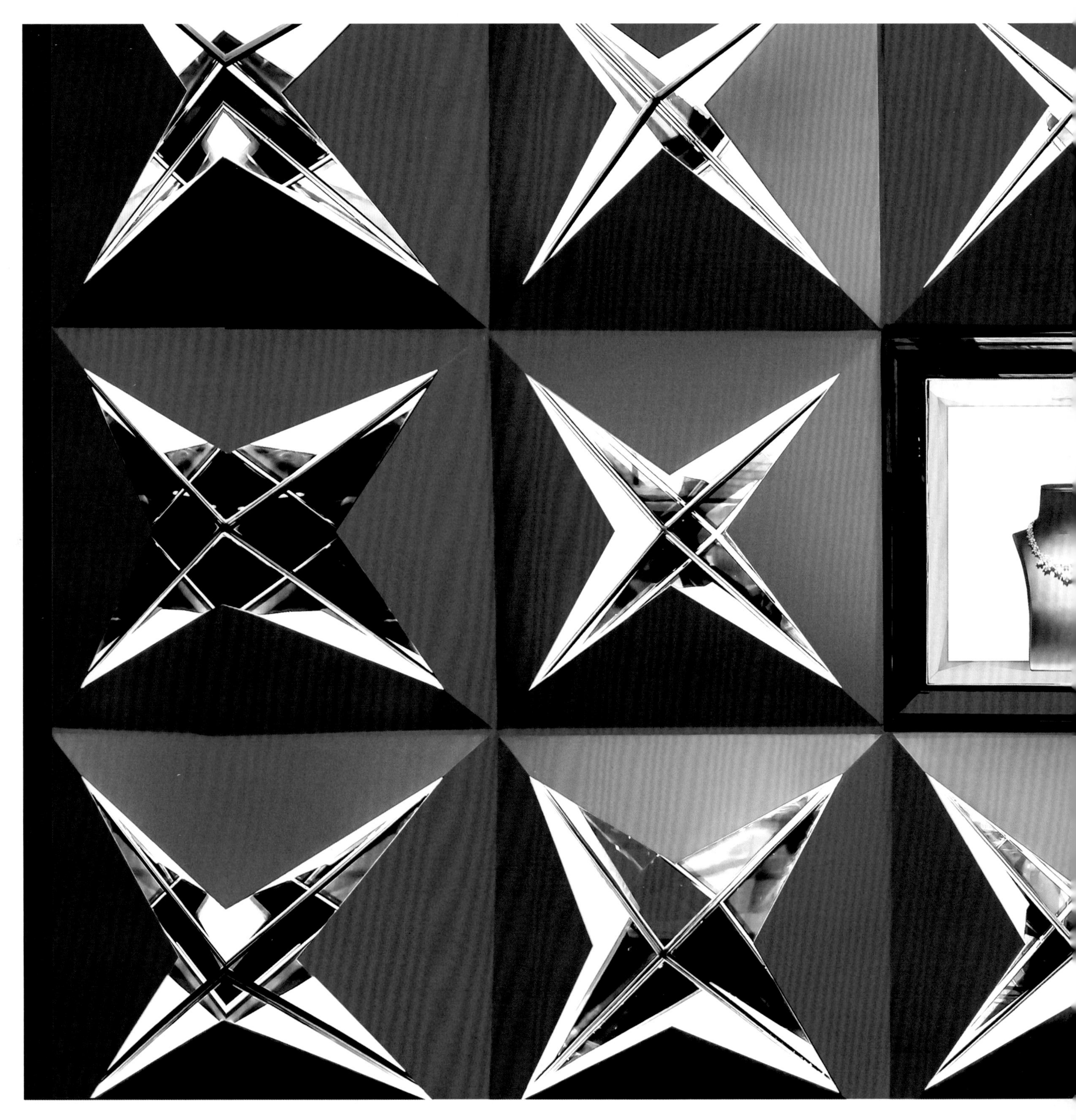

Wolfers

Wolfers

Wolfers

Wolfers

Fournisseur Breveté de la Cour

Wolfers

DERAIN
BY WOLFERS
Wolfers

Purple Lake International Golf

紫湖国际高尔夫

GOLF VILLA

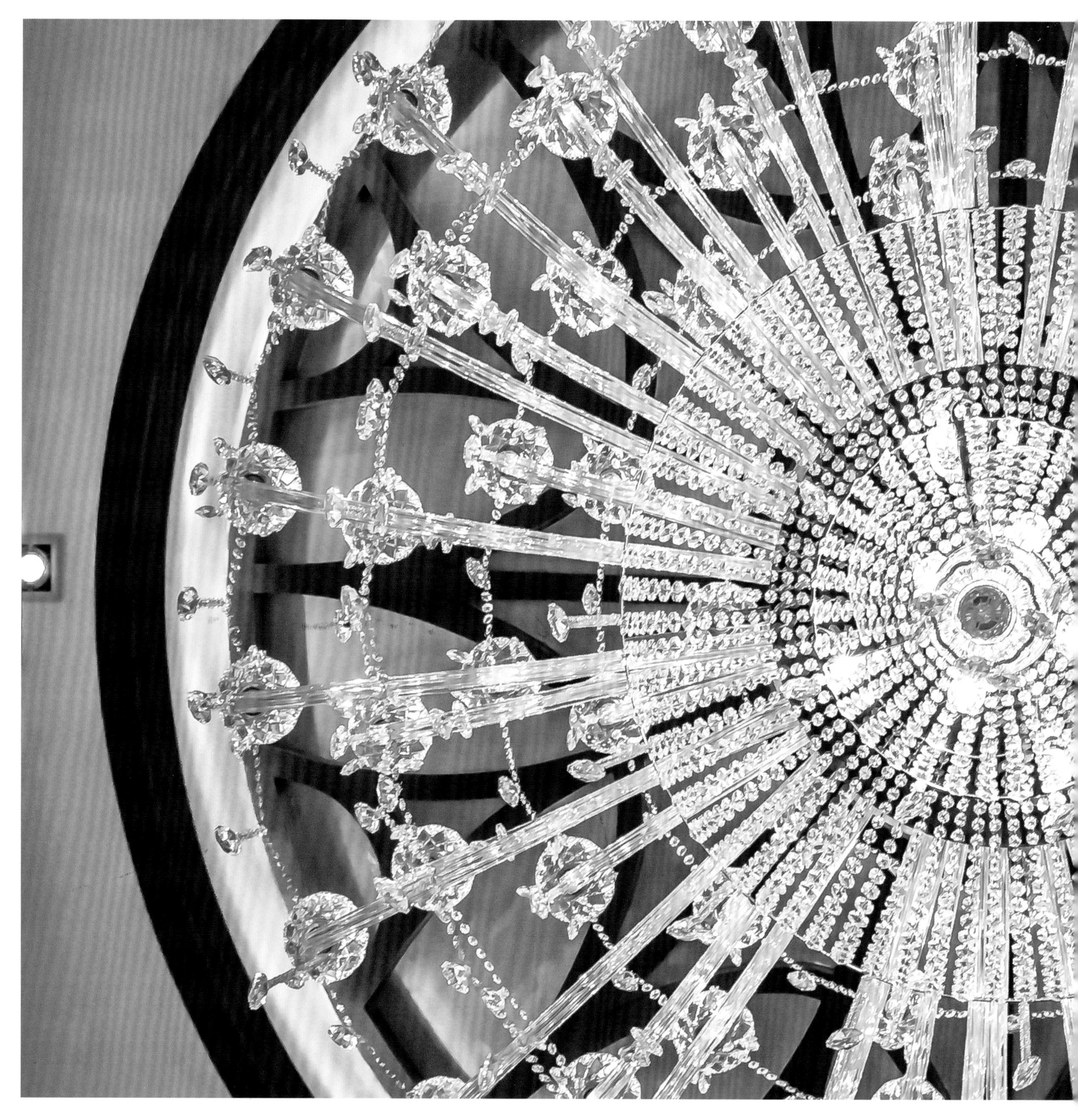

Ten Miles Riverside Sales Centre

碧桂园十里江湾

铂金
五星秘

The Glorious City

龙湖新壹城

大虹桥

龙湖新壹城
THE GLORIOUS CITY

Times City Sales Office

香港九龙仓时代上城

As a romantic country the national flowers of France, and then it so well, design and color is beautiful, in the group of flowers has a transcendental and free from vulgarity temperament. Legend has it the king of the kingdom of France's first dynasty Clovis in to accept baptism, god gave him a present, is the iris. In the design of the case, and then fruitful purple blue petals, large area of green pointed leaves, this as a foil of marigold yellow flower, flower the brick red soil, almost every color are designers carefully capture.

作为浪漫之国——法国的国花，鸢尾花花姿婀娜，花色美丽，在群花之中自有一种超凡脱俗的气质。相传法兰西王国第一个王朝的国王克洛维在接受洗礼时，上帝送给他一件礼物，就是鸢尾花。本案设计中鸢尾花丰硕的紫蓝色花瓣，大面积绿色的尖头长叶，作为衬托的万寿菊的黄色花蕊，花丛下砖红色的土壤，几乎每一处色彩都被设计师悉心捕捉。

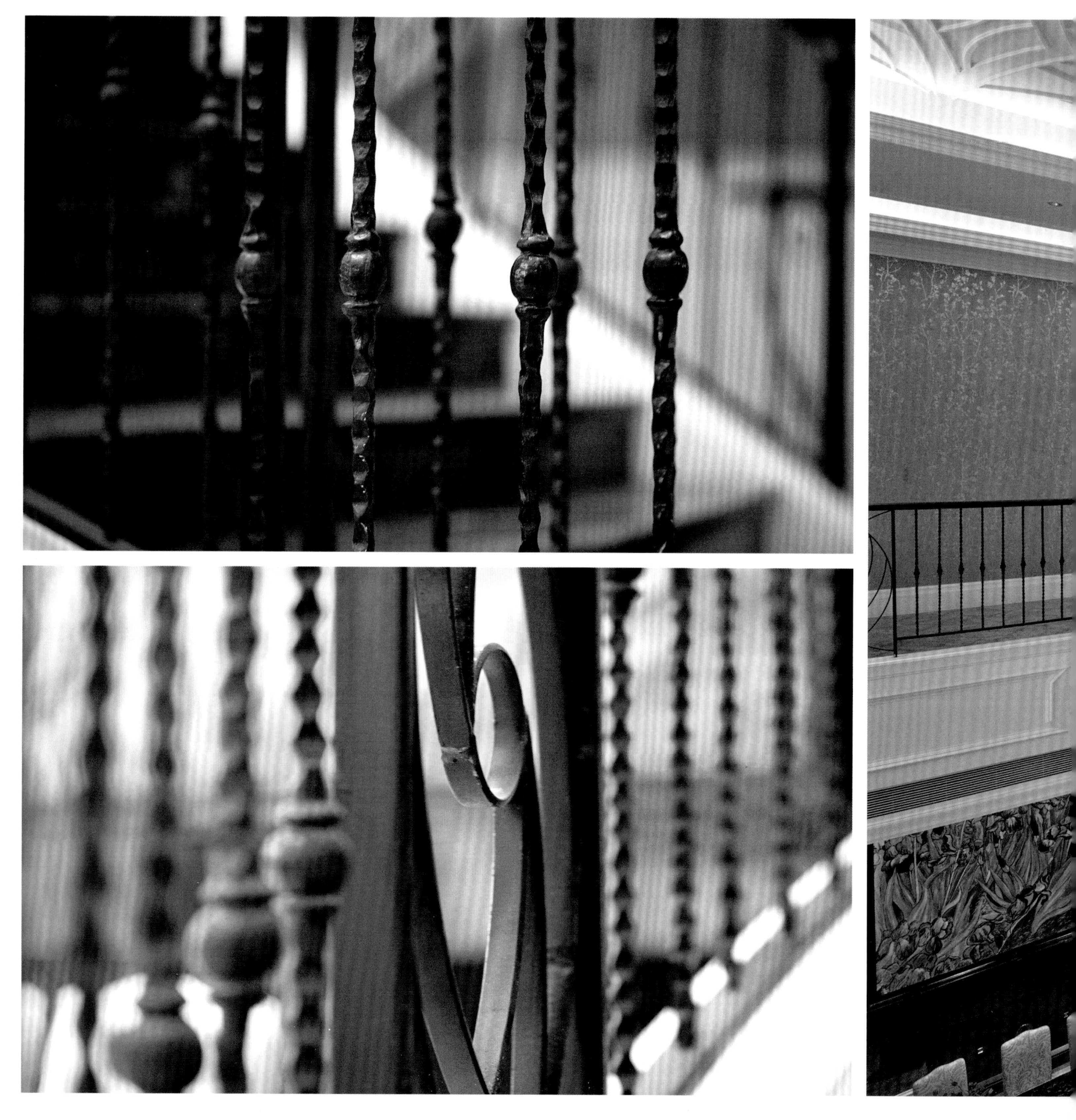

Times City
九龍倉|時代上城

Mingshihuating Sales Office

香港南益名士华庭

Strelitzia Reginae, also known as Bird of Paradise flower, is named for its shape similar to red-crowned crane raising its head and looking out. Its scientific name is given in honor of Queen Charlotte, the wife of King George III of England. The special shape and color of Strelitzia Reginae present good quality. Apparently this color looks peaceful, with some indistinct dynamism, and the red vitality is partly hidden and partly visible. In addition, it also includes the carefree element of orange, creating a fashionable and leisure aura.

鹤望兰，又称天堂鸟或极乐鸟花。以形似仙鹤昂首远望而得名，其学名是为纪念英国国王乔治三世王妃夏洛特而取的。鹤望兰花形奇特、色彩给人感觉很有质地。表面看上去这种色彩很安定，隐约之中却又有几分动感，红色的活力若隐若现。另外又包含了橙色轻松愉快的元素，营造出时尚且悠然自得的气氛。

Central Park Sales Centre

香港信和中央广场

Nob[illegible] conspicuous [illegible] eastern culture, the white lotus [illegible]med [illegible] flower of purity for its graceful and elegant temperament. The brilliant whiteness of this flower is an optimum choice for enhancing the overall brightness of a picture, presenting a clean and clear vision. Apart from its quietness, which is described in an ancient Chinese poem as "washed by clean waves while showing no coquet", white also displays conspicuous nobility for its brilliance and pureness.

白莲花，高雅脱俗，被喻为圣洁之花。白莲花的白色明度很高，这种色彩是提高整体画面明度的极佳选择，给人以洁净清澈的视觉效果。除了"濯清涟而不妖"的雅静，白色因其高明度和高纯度，彰显出耀眼、夺目的华贵气质！

天虹商场
摩尔莲花
购物广场

信和集团
Sino Group

中央广场
CENTRAL PARK

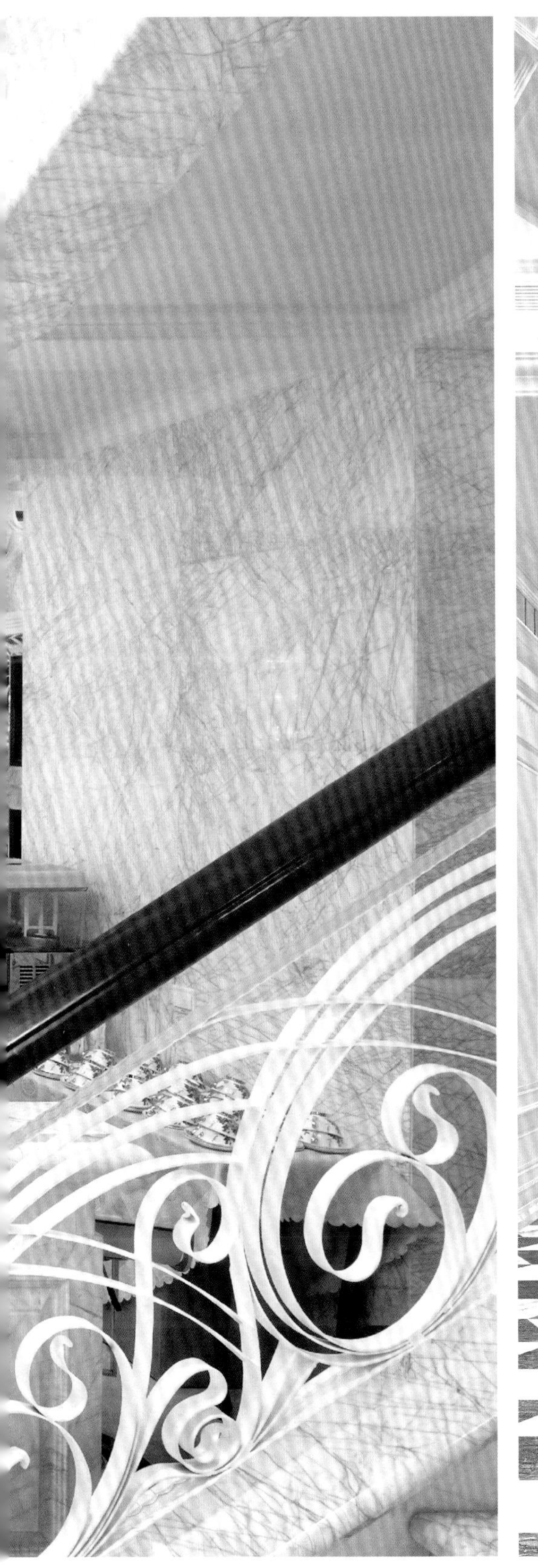

Langrun Sales Office

中能朗润国际

This is Langrun Sales Office, implanting the illusion of "butterflies" of uarious colors gradually flying towards all sides on the outer wall present fresh colors andmustery, attractingpeople before they enter the lobby. The dominant tone adopts the series of colors of blach-white-grey and gold-silver which has no hue, and it looks bright and elegant. Mild light is invisibly set behind each "butterfly" and the second ray of light set "butterflies" off to be more vivid and colorful.

朗润园销售会所将"蝴蝶谷"的梦幻植入现实。"蝴蝶领结"造型的大门、外墙上那些渐渐向四方飞去的各色"蝴蝶"，色彩清新优美而又具有某种神秘感，大厅的主色调选取了黑白灰和金银这一系列没有色相的色彩，明丽高雅，而每只"蝴蝶"下方并不显眼地安置了轻盈的提光，二次的光线将"蝴蝶"衬托得更加生动逼真且五彩缤纷。

The Century Sales Centre

中海世纪公馆

There is luxury in simplicity, and cheerfulness in quietness. The green is partly hidden and partly visible in the space, generating invisible attraction and affinity. The compact and orderly layout seems random and ornaments are delicate and unconventional. Each detail is worth thinking.

平实中带点华贵，沉静中又有些许轻快。绿意在空间中若隐若现，会所有种无形的吸引力、亲和力。布局紧凑有序，看似随意的摆件精致不落俗套，每一个细节都耐人寻味。

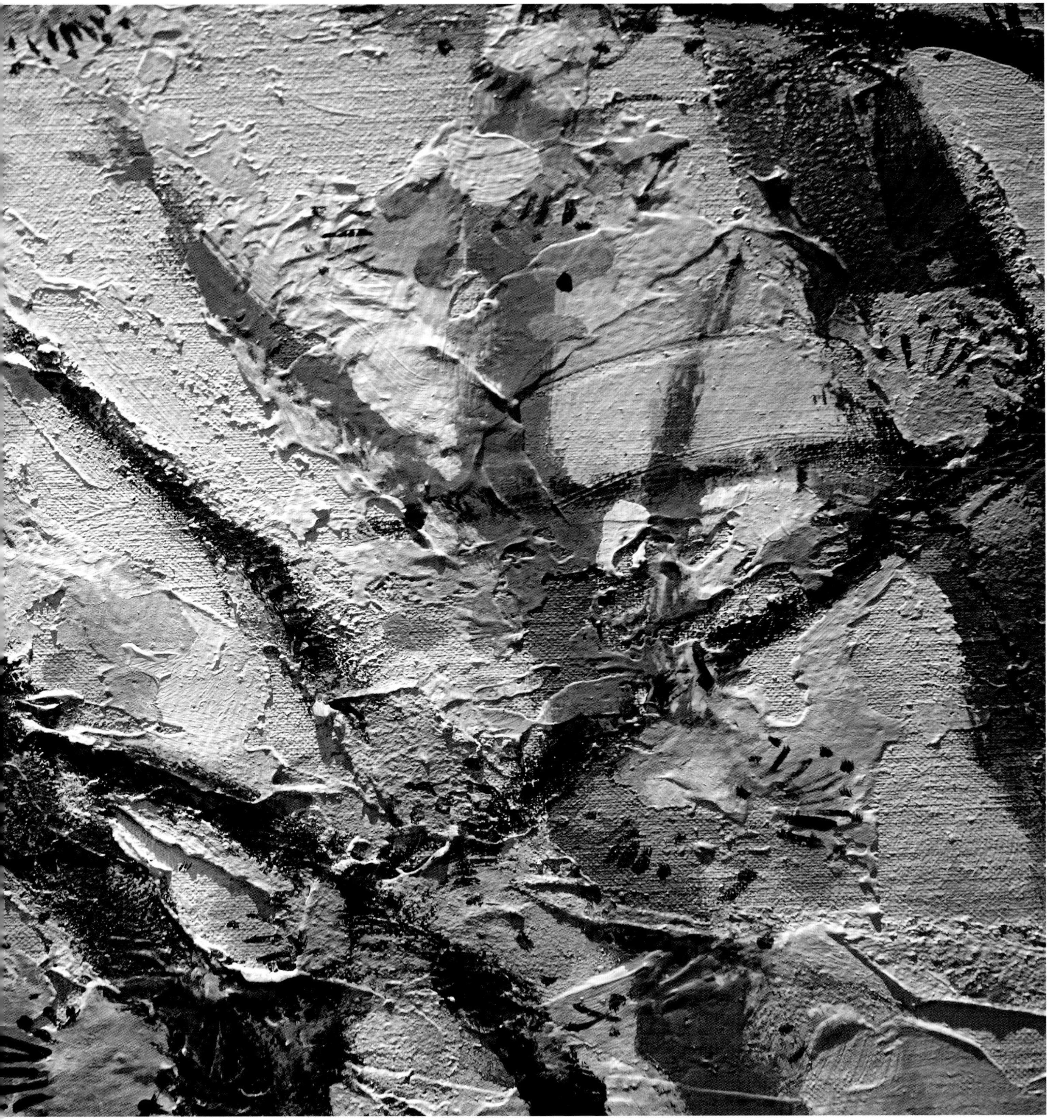

中海·世纪公馆销控表
中海地产
源自香港，工料之王
实力巨匠，誉满全球
中海地产
34年40城，240余项目
千亿央企，行业领军

中海·世纪公馆
THE CENTURY

中海·世纪公馆
THE CENTURY

Hunligong Club

华侨城婚礼宫

Bright and lofty :Cornflower The Cornflower has its home in the Europe, especially prevailing in Germany. The blue Cornflower bears a bright and clean color that resemble the sky, creating a feel of serenity and nobility, which would easily captivate bystanders. Compared with the warm colors, a member of the cold color family,blue is typically calm. Like green plants, blue colors have a composing and purifying effect on the mind. An impression of freshness and brightness follows the rising of brilliance and purity.

车矢菊，明净高远。车矢菊的故乡是欧洲，在德国几乎漫山遍野都是。蓝色的车矢菊有近乎天空般明净美丽的颜色，酿造出安详而崇高的感觉。这样的蓝色，会让人驻足仰视而不舍离去。与暖色系相比，冷色系中的蓝色往往给人冷静和安宁的感觉。蓝色系的色彩与绿色植物一样，具有镇静、净化心灵的功能，在提高其明度和纯度后，清爽、明朗的感觉也随之而生。

消火

Park Mansion Sales Office

龙湖世纪峰景

Rich and calm: pine cone brown While brown always implies solemnity, pine cone brown with a slight golden tint is able to make a breakthrough with a light softness and tenderness when displaying nobility, reminding one of the scenery of a distant mountain in the early fall, rich and tranquil, with a profound implication. In the solid color background, this color exhibits a strong and ready passion, especially s mart and elegant in response to cold gray.

松果褐，丰足安宁。尽管褐色总是呈现一种庄重而厚实的特点，带着些许浅金的松果褐却能从厚重的感觉中突围，作品在演绎华贵典雅的同时，平添了一份淡淡的柔和与温情，令人不禁怀想初秋远山的景致，意境深远悠长。在沉稳的色泽基调下，这种色彩透着一种蓄势待发的热情，尤其在与灰白等冷色系相融合时，它因温暖而尤显灵动优雅。

Crec Renhe Plaza

中铁仁禾广场

Hilfiger
Gieves
Lacoste

Burbe
B erryf

aha
果留
永辉超市
Bravo
YH

永辉超市
Bravo YH
水果
FRUITS
天天特价 新鲜健康
SALE
面点
NOODLES
熟食
FOODS
30.00

Custome
S.T.DUPONT
LANCOME
a Ricci
Lacoste
ORDANO
LANCOME

Gucci
soni
Lacoste
HILFIGER
L&SHARK

ERKE

UTICA
acoste

Tommy Hilfiger
ves&Hawkes
PANTENE
PRO-V

Yihe Mansion

中铁颐禾公馆

Lakeside Plaza

香港信和湖畔广场

Langrun International Hotel

中能朗润国际酒店

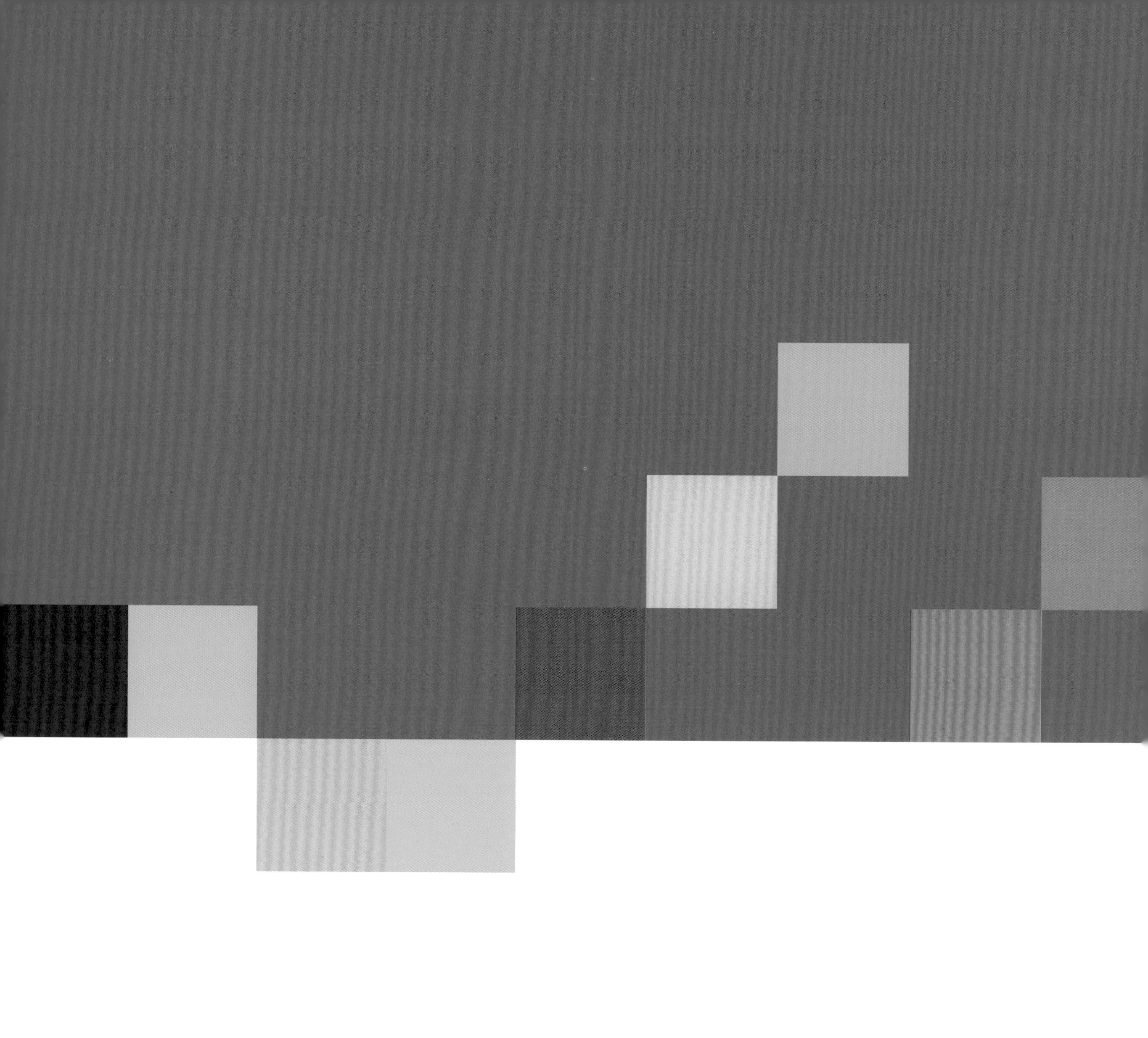

Macau MINI IN Hotel

澳门 MINI IN 精品酒店

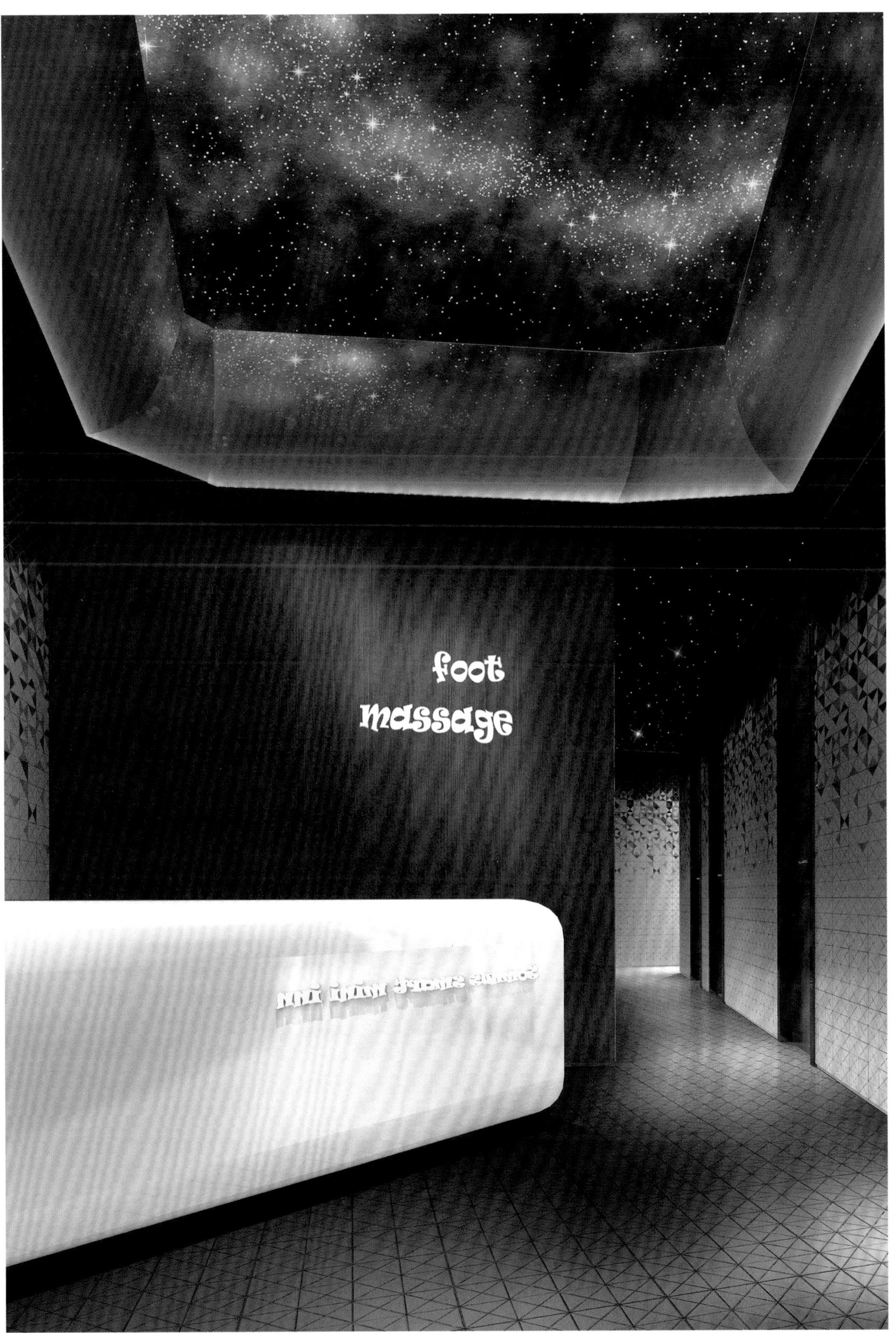
foot
massage

Happiness City Ⅰ

京鹏南悦城 Ⅰ

LOVE

Happiness City II

京鹏南悦城 II

YANKEE CAND
YANKEE CANDLE

La Cite Sales Centre

中海寰宇天下

Interio Design – Kinney Chan & Associates – FF&E – Photo – FongWong

中海·寰宇天下
LA CITÉ
千亿央企 全球500强
12年蝉联行业品牌价值第一
全球第14座寰宇天下

Acknowledgments 鸣谢

本书得以顺利面世，全赖各方的参与与支持，在此由衷感谢“香港方黄建筑师事务所”全体同仁的努力与付出！也衷心感谢以下合作伙伴对我司的大力支持和信任！

九龙仓（中国香港）
置地集团（中国香港）
中海地产
华侨城地产
华润地产
远航地产
中能置业
滕王阁地产
华人地产
信和（中国香港）
南益（中国香港）
龙湖地产
绿地集团
珠江地产
远大地产
大华地产
光达地产
新希望地产
新榕建筑置业（中国澳门）
新建利建筑置业（中国澳门）
碧桂园地产
中铁集团
中粮地产
世茂地产
长航地产
恒河地产
恒丰地产

香港方黄建筑师事务所
于 1997 年在中国香港创立，是专业提供多元建筑
居住、商业空间整体设计与统筹管理的营运机构。事务所经过近二十年的发展，
从香港到深圳、上海、成都相继成立了高端专业的国际化设计与管理服务
团队，团队至今拥有中国、意大利、加拿大等多个地区的优秀合伙人与设计师。事务所
以不断创新务实的设计理念及服务赢得了众多客户的支持与信赖。
合作客户既有香港、澳门等地的知名房地产开发商，
如九龙仓、信和、置地、南益地产、新建利建筑置业……也有内地大型房地产开发企业，
如中海、龙湖、碧桂园、世茂以及中粮、华润、绿地、中铁、华侨城地产等。项目作品已遍布纽约、香港、澳门、
北京、上海、广州、深圳、厦门、天津、成都、重庆、沈阳等多个城市及地区。

Hong Kong Fong & Wong Architects & Associates
was established in Hong Kong in 1997, which specializes in integrated design and management of diversified residential and commercial space. After development for over twenty years, besides the head office, Fong & Wong has established high-end professional and international design and anagement service teams in Shenzhen, Shanghai and Chengdu. Fong & Wong has attracted excellent partners and designers from many countries and areas including China Mainland, Hong Kong SAR, Italy and Canada. By virtue of the design philosophy of constant innovation and pragmatic services, Fong & Wong has won a number of customers' support and trust.
The cooperation customers cover both well-known Hong Kong property developers including Wharf, Sino Group, Hong Kong Land, South Asia Real Estate, and Newly Built Construction Property, etc. and famous mainland property developers including Zhonghai, Longfor, Country Garden, ShiMao and COFCO, China Resources Land, Greenland, China Railway Construction Real Estate Group, and OTC etc.
Its design projects and works can be seen all over New York, Hong Kong, Macao, Beijing, Shanghai, Guangzhou, Shenzhen, Xiamen, Tianjin, Chengdu, Chongqing, Shenyang and many other cities and areas.

图书在版编目（CIP）数据

设计有约 3 / 方峻 著 . – 武汉 : 华中科技大学出版社 , 2016.11

ISBN 978-7-5680-2260-6

Ⅰ . ①设… Ⅱ . ①方… Ⅲ . ①室内装饰设计 Ⅳ . ① TU238

中国版本图书馆 CIP 数据核字（2016）第 241636 号

设计有约 3
Sheji Youyue 3

方峻 著

出版发行：华中科技大学出版社（中国·武汉） 电话：（027）81321913
武汉市东湖新技术开发区华工科技园 邮编：430223

责任编辑：熊纯 特邀编辑：董莉婷 排版设计：筑美文化
责任校对：赵营涛 封面设计：王伟 责任监印：张贵君

印 刷：中华商务联合印刷（广东）有限公司
开 本：889 mm × 1194 mm 1/12
印 张：18.5
字 数：111 千字
版 次：2016 年 11 月第 1 版 第 1 次印刷
定 价：278.00 元（USD 55.99）

投稿热线 : 13710226636 duanyy@hustp.com
本书若有印装质量问题，请向出版社营销中心调换
全国免费服务热线：400-6679-118 竭诚为您服务